1 2 3 4 5 6 7 8 9 10

Draw a line to match each group of things to its correct number.

Draw a line to match each group of things to its correct number.

11

12

13

14

15

16

17

18

19

20

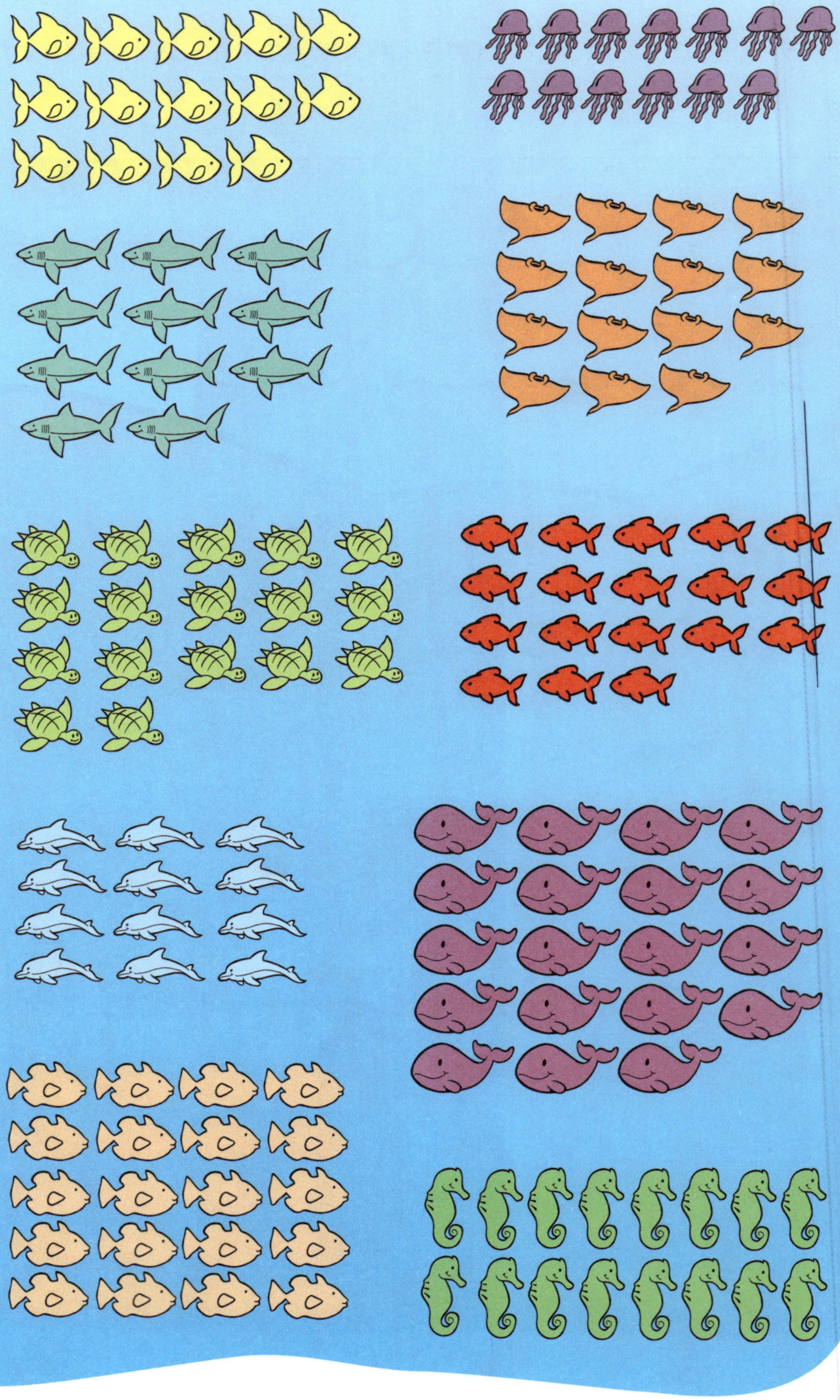

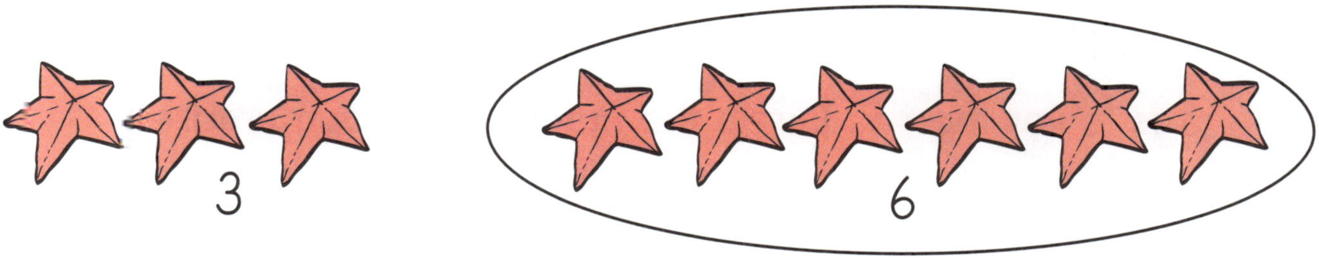

Greater than means more than. 6 is greater than 3. 6 is more than 3. Circle the group that is greater.

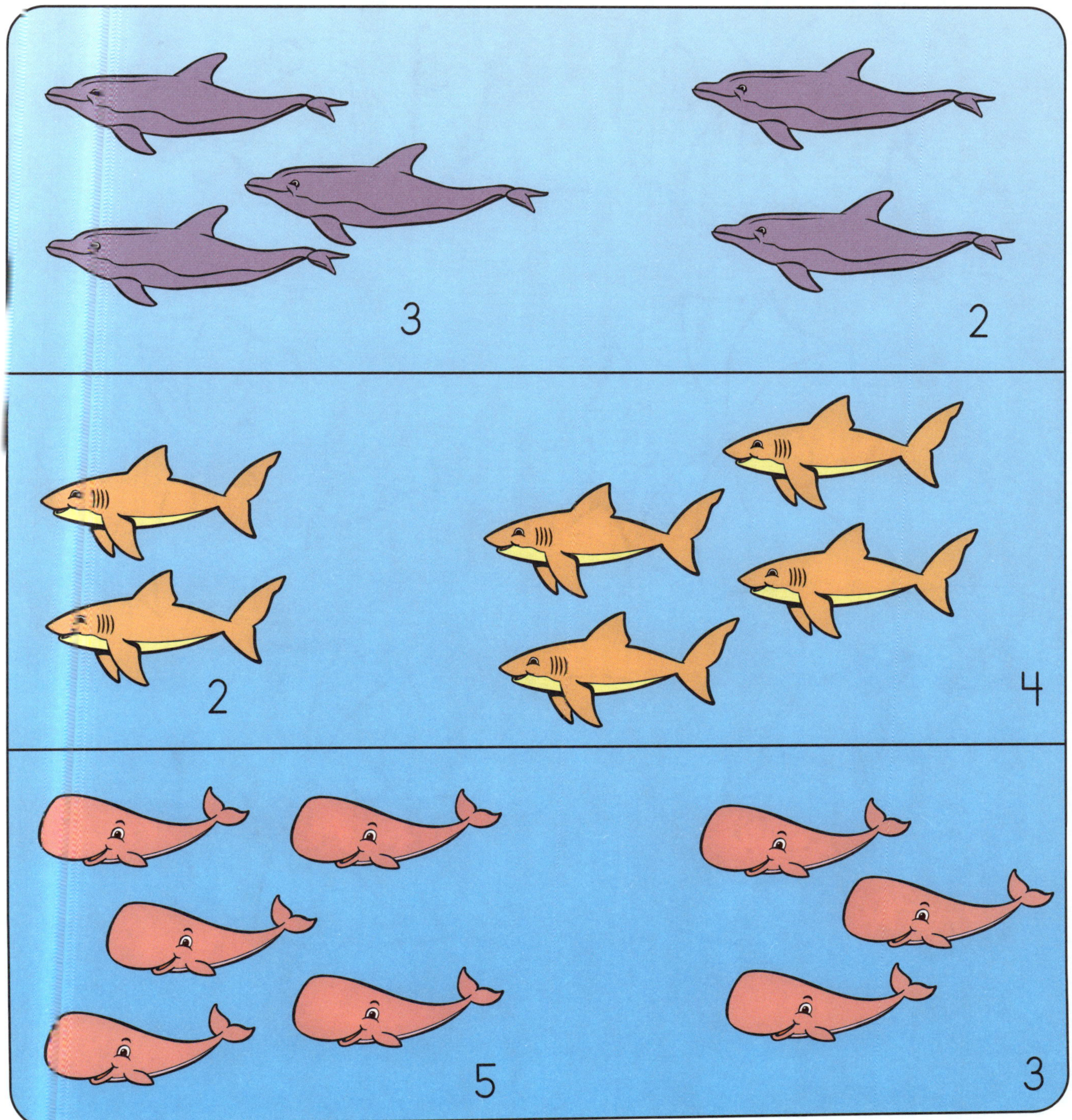

Less than means fewer than. 5 is less than 9. 5 is fewer than 9. Circle the number on each umbrella that is less than the other number.

Connect the numbers from 1 to 20
to find something you can make at the beach.

1 2 3 4 5 6 7 8 9 10 11 12 13 14 15 16 17 18 19 20

Count by 2s. Write the missing numbers.

1		3		5
	7		9	
11		13		15
	17		19	

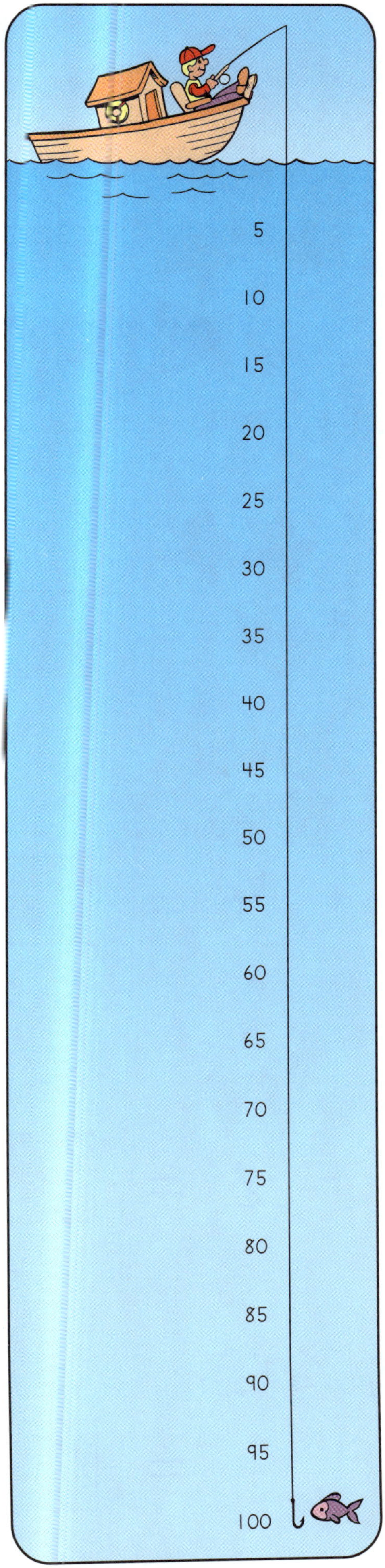

To count by 5s, say every number in order that ends in 0 or 5. Connect the numbers in order to find something that swims in the ocean.

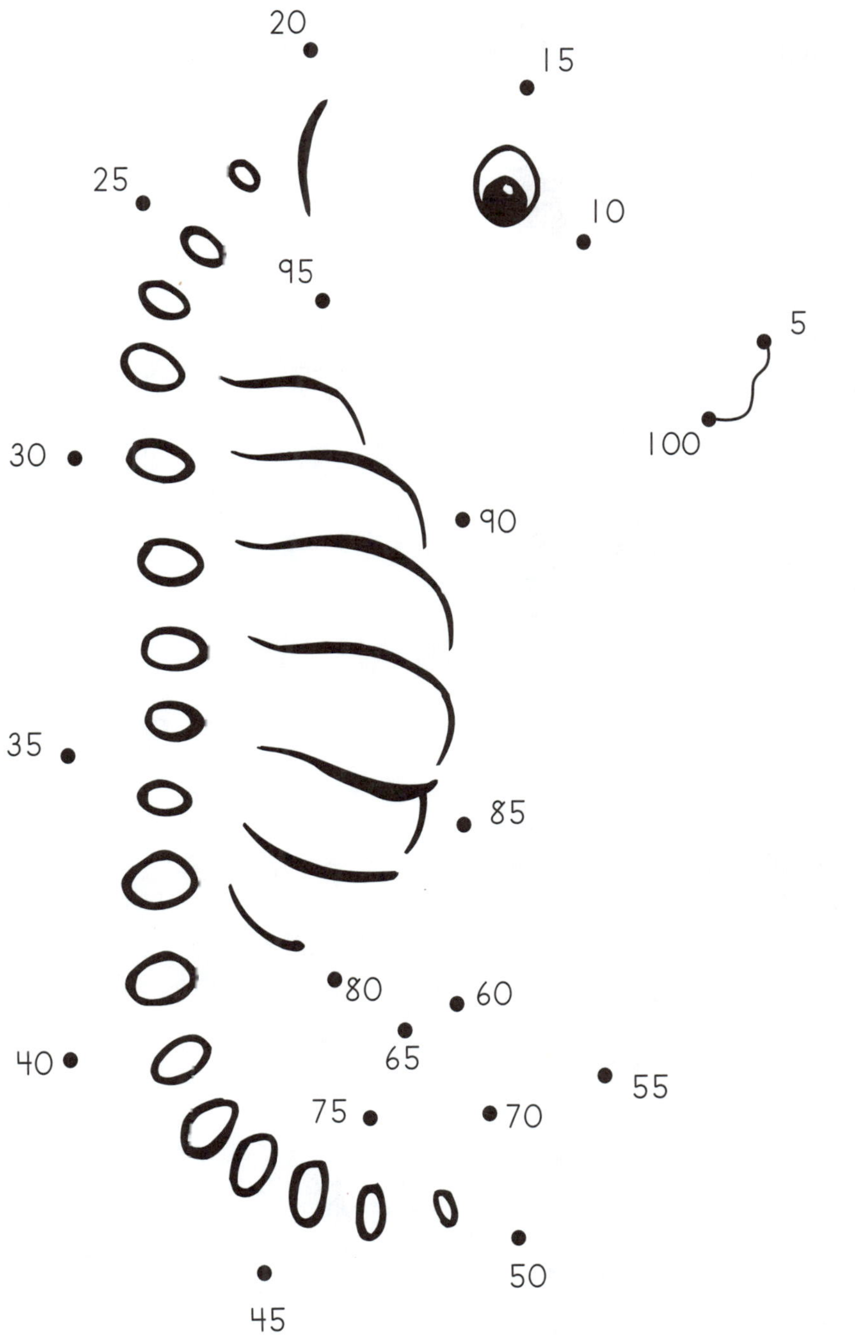

1 + 3 = 4

Count the things in each group. Add the two numbers together and write the answers in the boxes.

1 + 1 = ☐

1 + 2 = ☐

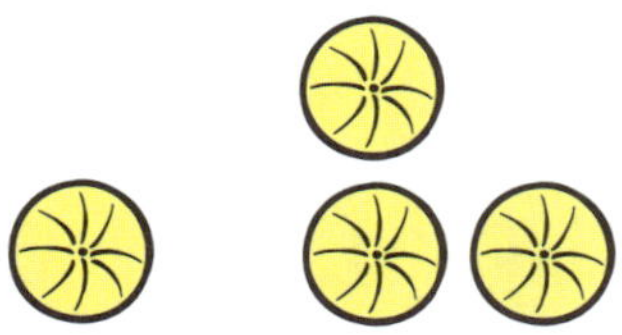

1 + 3 = ☐

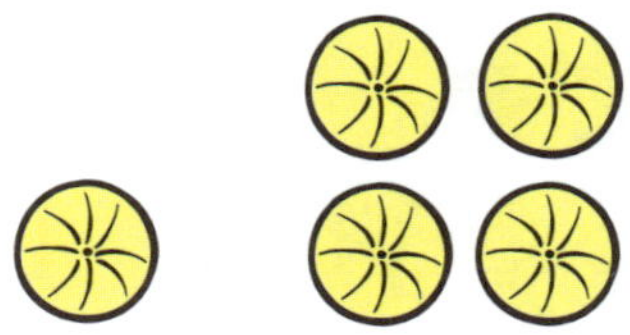

1 + 4 = ☐

2 + 1 = ☐

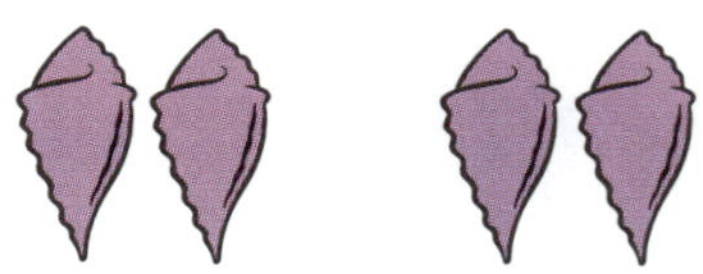

2 + 2 = ☐

2 + 3 = ☐

3 + 1 = ☐

Add and write the sums.

$$\begin{array}{r} 1 \\ +1 \\ \hline \square \end{array} \qquad \begin{array}{r} 1 \\ +2 \\ \hline \square \end{array} \qquad \begin{array}{r} 1 \\ +3 \\ \hline \square \end{array}$$

$$\begin{array}{r} 2 \\ +1 \\ \hline \square \end{array} \qquad \begin{array}{r} 2 \\ +2 \\ \hline \square \end{array} \qquad \begin{array}{r} 2 \\ +3 \\ \hline \square \end{array}$$

$$\begin{array}{r} 3 \\ +1 \\ \hline \square \end{array} \qquad \begin{array}{r} 3 \\ +2 \\ \hline \square \end{array} \qquad \begin{array}{r} 3 \\ +3 \\ \hline \square \end{array}$$

$4 - 2 = \boxed{2}$

Count all the things in each group. Then count how many are crossed out and write the difference.

$2 - 1 = \square$

$3 - 2 = \square$

$3 - 1 = \square$

$4 - 3 = \square$

$4 - 1 = \square$

$4 - 2 = \square$

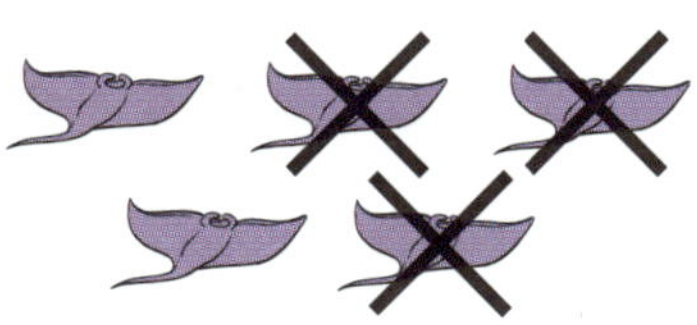

$5 - 3 = \square$

$5 - 4 = \square$

$$\begin{array}{r}5\\-2\\\hline \boxed{3}\end{array}$$

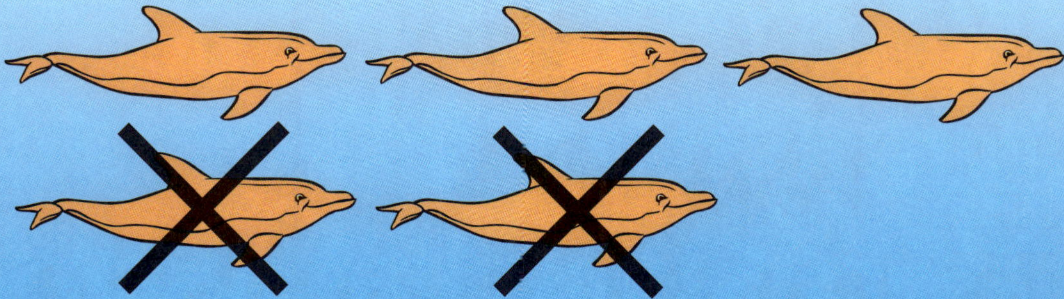

• Check your answers!

Subtract and write the differences.

$$\begin{array}{r}5\\-1\\\hline \boxed{}\end{array}$$

$$\begin{array}{r}3\\-2\\\hline \boxed{}\end{array}$$

$$\begin{array}{r}4\\-3\\\hline \boxed{}\end{array}$$

$$\begin{array}{r}2\\-1\\\hline \boxed{}\end{array}$$

$$\begin{array}{r}5\\-4\\\hline \boxed{}\end{array}$$

$$\begin{array}{r}4\\-2\\\hline \boxed{}\end{array}$$

$$\begin{array}{r}4\\-3\\\hline \boxed{}\end{array}$$

$$\begin{array}{r}3\\-1\\\hline \boxed{}\end{array}$$

$$\begin{array}{r}5\\-2\\\hline \boxed{}\end{array}$$

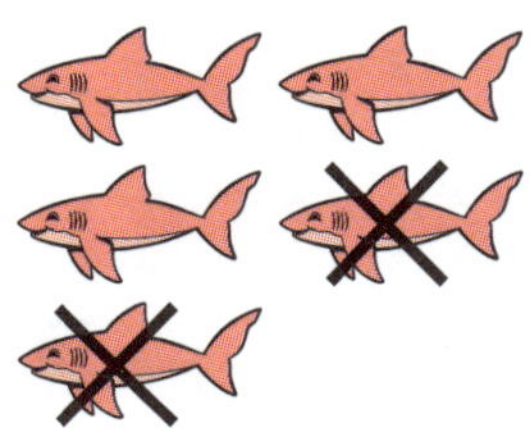

$7 + 5 = \boxed{12}$

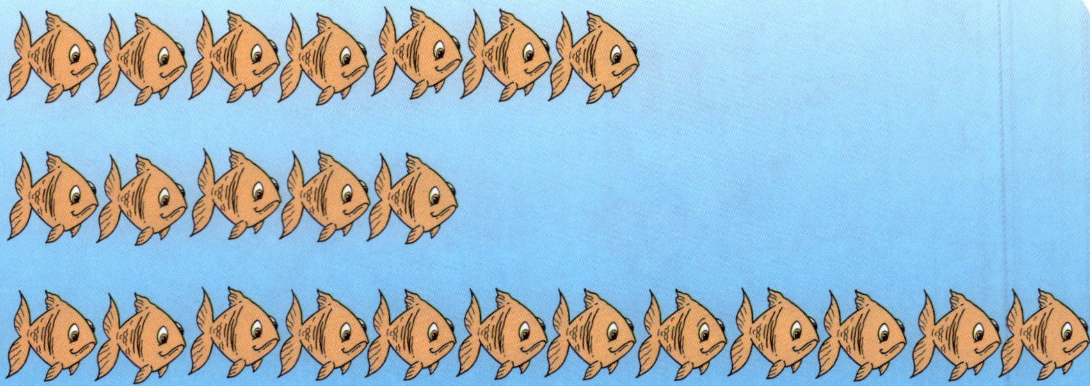

Add and write the sums.

$5 + 4 = \square$ $7 + 1 = \square$ $8 + 3 = \square$ $6 + 2 = \square$

$7 + 4 = \square$

$9 + 2 = \square$

$6 + 3 = \square$ $5 + 5 = \square$ $9 + 3 = \square$ $8 + 2 = \square$

$6 + 4 = \square$ $8 + 2 = \square$ $9 + 1 = \square$ $5 + 2 = \square$

$6 + 6 = \square$

$\begin{array}{r} 9 \\ -3 \\ \hline 6 \end{array}$

Subtract and write the differences.

$\begin{array}{r} 12 \\ -1 \\ \hline \end{array}$	$\begin{array}{r} 3 \\ -2 \\ \hline \end{array}$	$\begin{array}{r} 9 \\ -1 \\ \hline \end{array}$	$\begin{array}{r} 8 \\ -2 \\ \hline \end{array}$		$\begin{array}{r} 6 \\ -2 \\ \hline \end{array}$
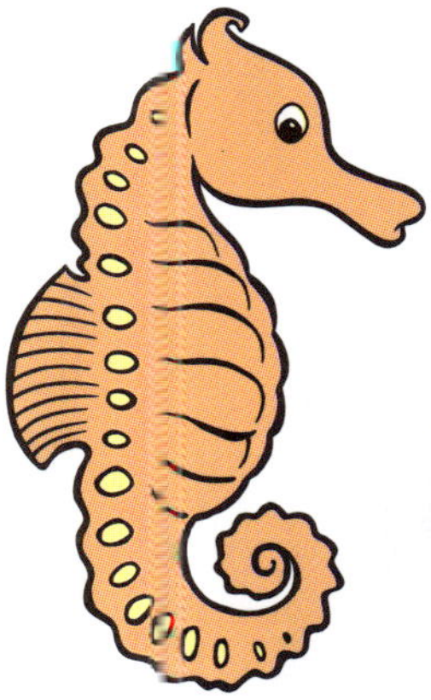	$\begin{array}{r} 7 \\ -4 \\ \hline \end{array}$	$\begin{array}{r} 11 \\ -3 \\ \hline \end{array}$	$\begin{array}{r} 10 \\ -6 \\ \hline \end{array}$	$\begin{array}{r} 4 \\ -2 \\ \hline \end{array}$	$\begin{array}{r} 5 \\ -1 \\ \hline \end{array}$
$\begin{array}{r} 5 \\ -2 \\ \hline \end{array}$	$\begin{array}{r} 6 \\ -5 \\ \hline \end{array}$	$\begin{array}{r} 9 \\ -6 \\ \hline \end{array}$		$\begin{array}{r} 7 \\ -2 \\ \hline \end{array}$	$\begin{array}{r} 4 \\ -1 \\ \hline \end{array}$

Watch the sign! Add or subtract to solve each problem.
Write the answers.

$5 + 1 = \square$

$7 + 5 = \square$

$12 - 6 = \square$

$5 + 3 = \square$

$12 - 3 = \square$

$11 - 7 = \square$

$5 + 7 = \square$

$10 - 4 = \square$

$8 + 4 = \square$

$6 + 5 = \square$

$6 + 1 = \square$

$7 + 3 = \square$

$11 - 8 = \square$

$5 + 2 = \square$

$10 - 2 = \square$

Circle each group of 10 things.
Write the number that tells how many tens.

How many tens?

How many tens?

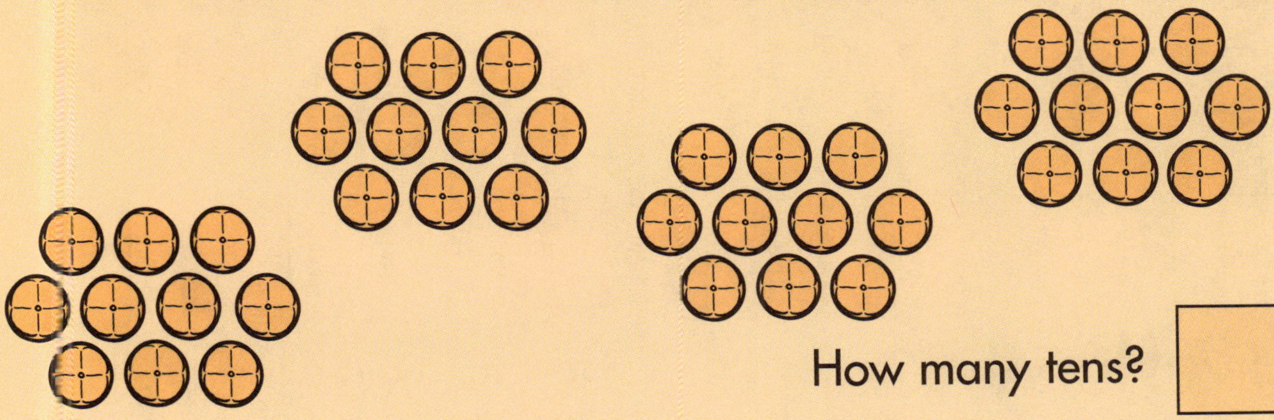

How many tens?

Count and circle each group of 10 things.
Write how many tens and how many ones.

☐ ten + ☐ one

How many in all? ☐

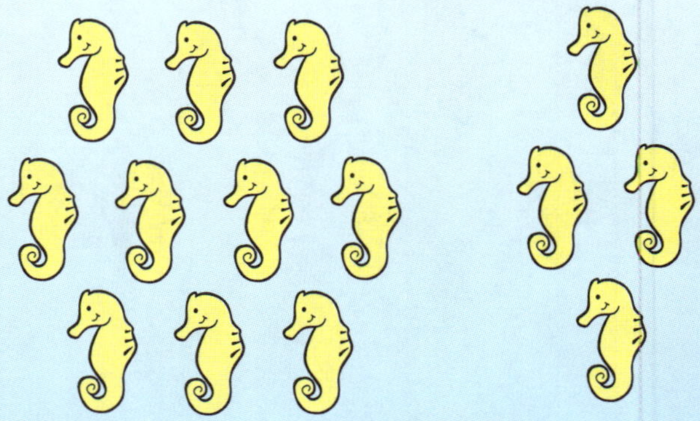

☐ ten + ☐ ones

How many in all? ☐

☐ ten + ☐ ones

How many in all? ☐

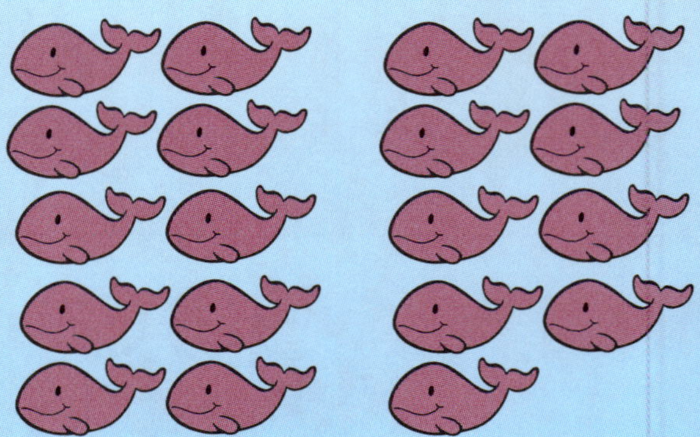

☐ ten + ☐ ones

How many in all? ☐

Add the ones and write the sums. Then add the tens and write the sums. How many in all?

21	13	35	47	52
+35	+32	+42	+31	+17

48	66
+41	+22

74	83	25	38	56
+23	+16	+53	+51	+43

Subtract the ones and write the differences.
Then subtract the tens and write the differences.
How many are left?

44	57	34	26	28
−23	−31	−21	−14	−11

37	65	77	89	93
−15	−32	−44	−57	−21

To add three numbers, add the first two numbers together. Then add the sum to the last number. Write the answers below.

1 +3=4 + 4	2 +2=4 + 4	3 +3=6 + 6	4 +1=5 + 2
5 +2=7 + 2	2 +4=6 + 2	1 +1=2 + 6	3 +5=8 + 1

Add the first two numbers. Then add the bottom number to your answer. Write the sums.

2	2	3	4	1
1	3	4	1	4
+2	+3	+2	+2	+4
☐	☐	☐	☐	☐

3	2	5	4	5
3	2	3	4	4
+2	+5	+4	+4	+6
☐	☐	☐	☐	☐

Add or subtract to solve the problems.
Write the answers.

23	35	61	55	45
− 12	+ 42	+ 4	− 22	− 23

75	32	86	44	72
− 34	+ 17	− 25	− 13	+ 27

Write the missing numbers.

$5 + 1 = \boxed{6}$

$11 - \boxed{4} = 7$

$\boxed{} + 5 = 12$

$12 - \boxed{} = 6$

$5 + \boxed{} = 12$

$10 - 6 = \boxed{}$

$8 + \boxed{} = 10$

$\boxed{} - 4 = 2$

$4 + 5 + 2 = \boxed{}$

$6 + 2 + \boxed{} = 13$

$\boxed{} + 5 + 5 = 12$

$5 + \boxed{} + 7 = 15$

Measure each thing.

Circle the correct length on the ruler.

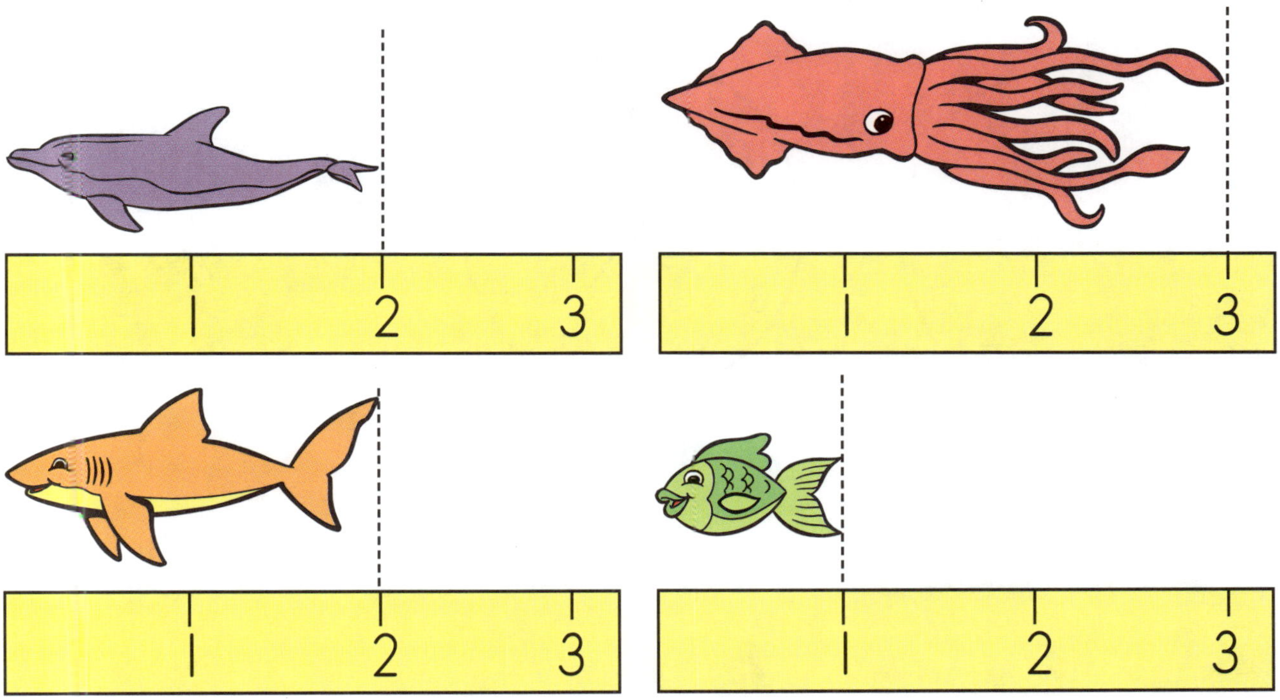

Circle the thing that is the longest.

The short hand tells the hour. Read each clock. Write the time.

____ : ____ ____ : ____ ____ : ____

____ : ____ ____ : ____ ____ : ____

The long hand tells the minutes. Read each clock.
Write the time.

___ : ___ ___ : ___ ___ : ___

___ : ___ ___ : ___ ___ : ___

Count pennies by 1s.

Count nickels by 5s.

Count dimes by 10s.

Count quarters by 25s.

Count the coins and write the amounts.

¢

¢

¢

¢

Circle the coins you need to buy each thing.

Read each story problem. Write the answer.

Sam found 5 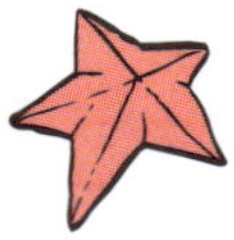 s. Corey found 7  s.

How many  s did they have in all?

Number Sentence: ____________________

Tonika had 12  s. She gave Tracy 4.

How many did she have left?

Number Sentence: ____________________

Bill made 8  s. Jim made 9  s.

How many did they make in all?

Number Sentence: ____________________

Ryan caught 19  .

4 got away. How many did he have left?

Number Sentence: ____________________

Answer Page

page 1

page 2

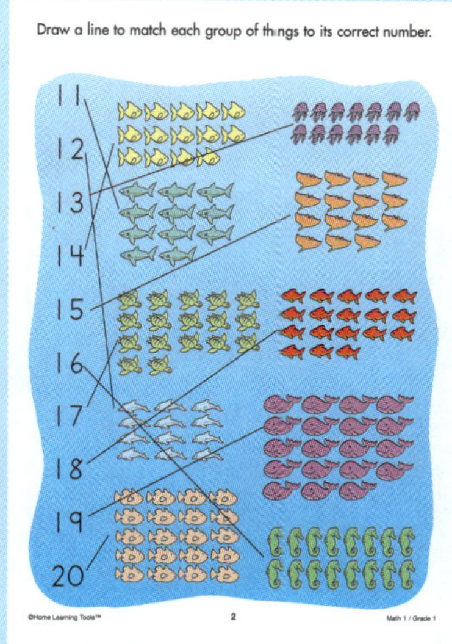

page 3

page 4

page 5

page 6

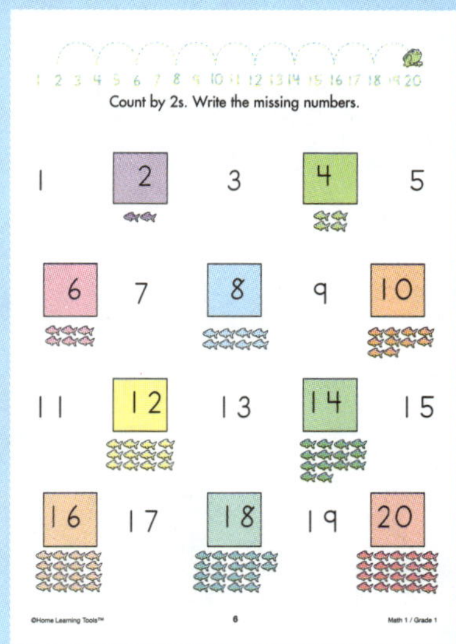

page 7

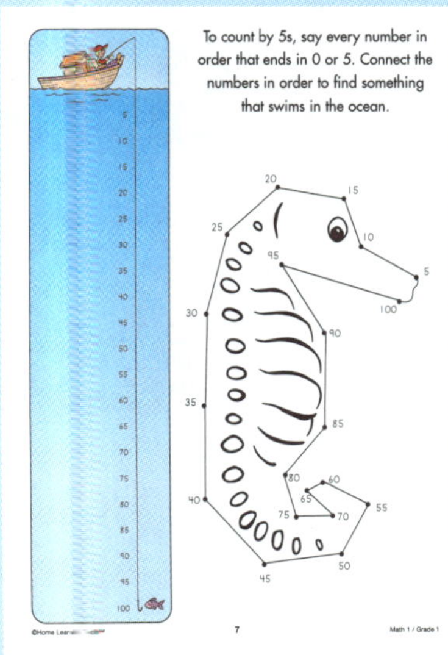

page 8

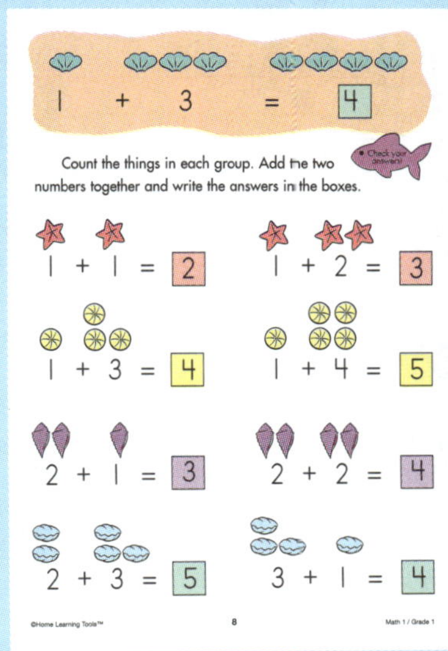

page 9

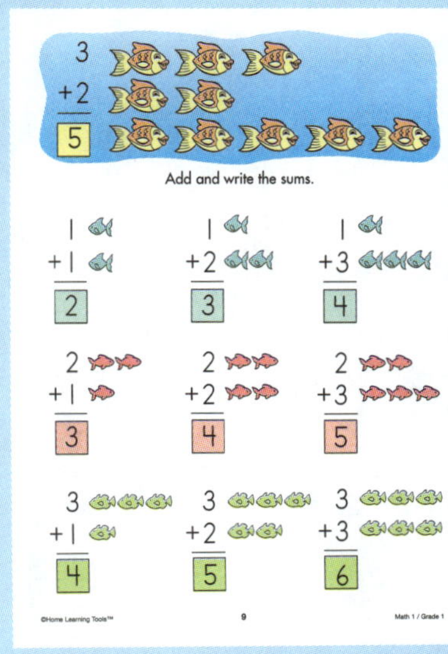

Answer Page

page 10

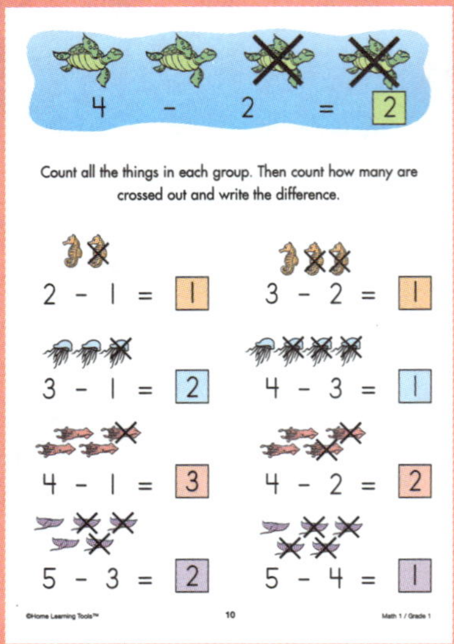

4 − 2 = 2

Count all the things in each group. Then count how many are crossed out and write the difference.

2 − 1 = 1
3 − 2 = 1
3 − 1 = 2
4 − 3 = 1
4 − 1 = 3
4 − 2 = 2
5 − 3 = 2
5 − 4 = 1

page 11

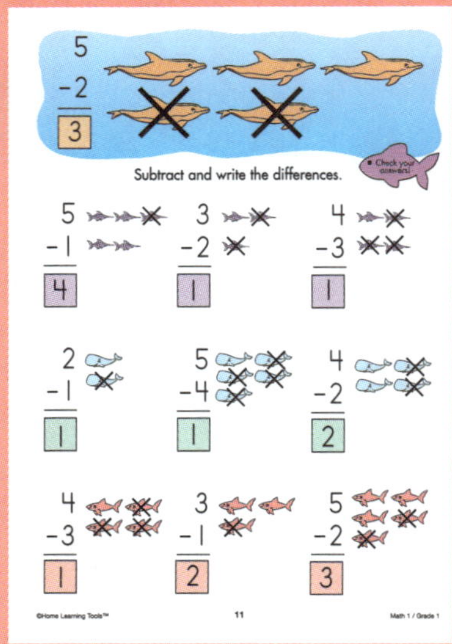

5 − 2 = 3

Subtract and write the differences.

5 − 1 = 4
3 − 2 = 1
4 − 3 = 1
2 − 1 = 1
5 − 4 = 1
4 − 2 = 2
4 − 3 = 1
3 − 1 = 2
5 − 2 = 3

page 12

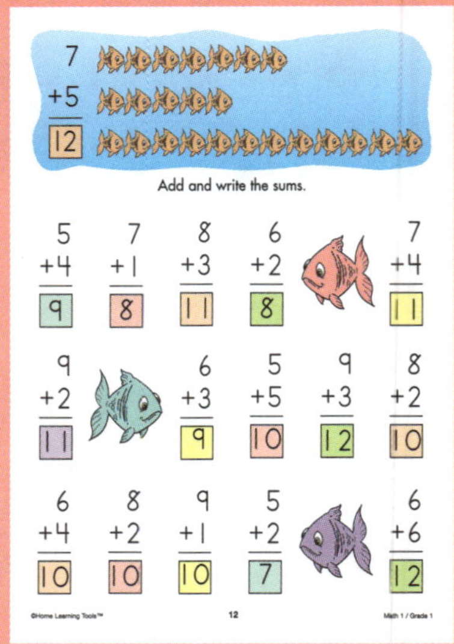

7 + 5 = 12

Add and write the sums.

5 + 4 = 9
7 + 1 = 8
8 + 3 = 11
6 + 2 = 8
7 + 4 = 11
9 + 2 = 11
6 + 3 = 9
5 + 5 = 10
9 + 3 = 12
8 + 2 = 10
6 + 4 = 10
8 + 2 = 10
9 + 1 = 10
5 + 2 = 7
6 + 6 = 12

page 13

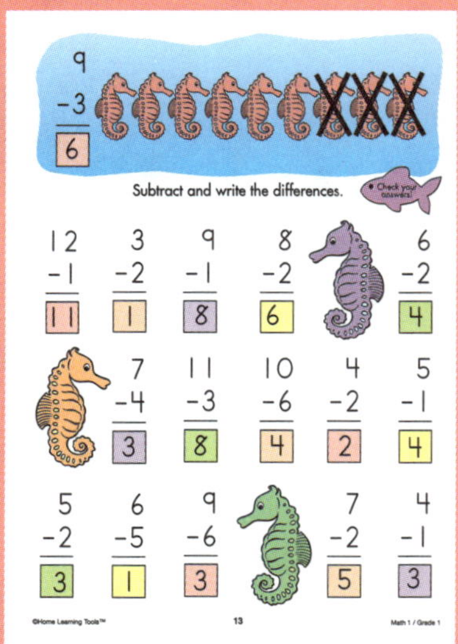

9 − 3 = 6

Subtract and write the differences.

12 − 1 = 11
3 − 2 = 1
9 − 1 = 8
8 − 2 = 6
6 − 2 = 4
7 − 4 = 3
11 − 3 = 8
10 − 6 = 4
4 − 2 = 2
5 − 1 = 4
5 − 2 = 3
6 − 5 = 1
9 − 6 = 3
7 − 2 = 5
4 − 1 = 3

page 14

Watch the sign! Add or subtract to solve each problem. Write the answers.

5 + 1 = 6
7 + 5 = 12
12 − 6 = 6
5 + 3 = 8
12 − 3 = 9
11 − 7 = 4
5 + 7 = 12
10 − 4 = 6
8 + 4 = 12
6 + 5 = 11
6 + 1 = 7
7 + 3 = 10
11 − 8 = 3
5 + 2 = 7
10 − 2 = 8

page 15

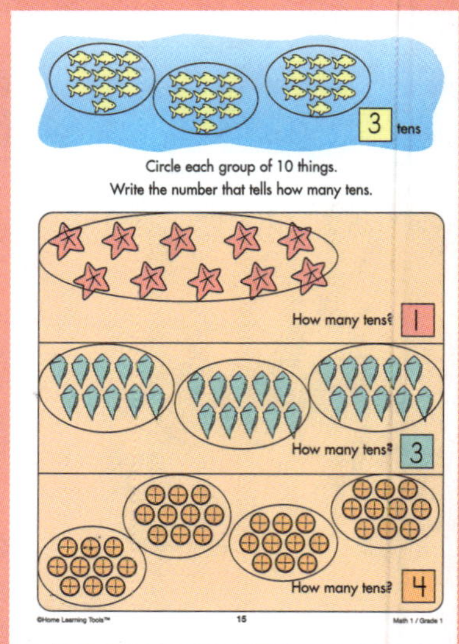

3 tens

Circle each group of 10 things. Write the number that tells how many tens.

How many tens? 1
How many tens? 3
How many tens? 4

page 16

1 ten + 2 ones

Count and circle each group of 10 things. Write how many tens and how many ones.

1 ten + 1 one. How many in all? 11
1 ten + 4 ones. How many in all? 14
1 ten + 7 ones. How many in all? 17
1 ten + 9 ones. How many in all? 19

page 17

21 + 34 = 55
14 + 61 = 75
42 + 53 = 95

Add the ones and write the sums. Then add the tens and write the sums. How many in all?

21 + 35 = 56
13 + 32 = 45
35 + 42 = 77
47 + 31 = 78
52 + 17 = 69
48 + 41 = 89
66 + 22 = 88
74 + 23 = 97
83 + 16 = 99
25 + 53 = 78
38 + 51 = 89
56 + 43 = 99

page 18

25 − 11 = 14

Subtract the ones and write the differences. Then subtract the tens and write the differences. How many are left?

44 − 23 = 21
57 − 31 = 26
34 − 21 = 13
26 − 14 = 12
28 − 11 = 17
37 − 15 = 22
65 − 32 = 33
77 − 44 = 33
89 − 57 = 32
93 − 21 = 72

Answer Page

page 19

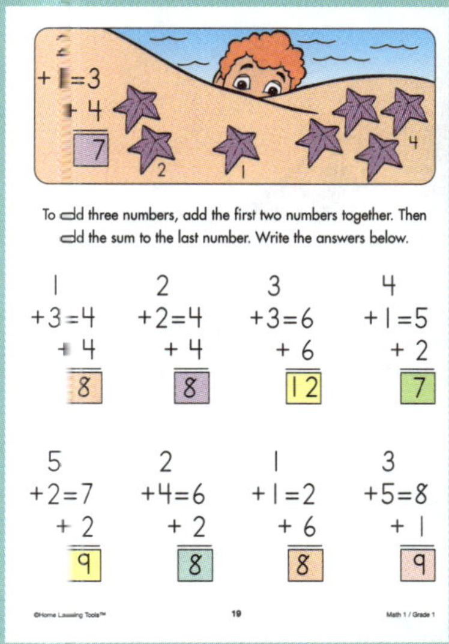
To add three numbers, add the first two numbers together. Then add the sum to the last number. Write the answers below.

1 +3=4 +4 = 8	2 +2=4 +4 = 8	3 +3=6 +6 = 12	4 +1=5 +2 = 7
5 +2=7 +2 = 9	2 +4=6 +2 = 8	1 +1=2 +6 = 8	3 +5=8 +1 = 9

page 20

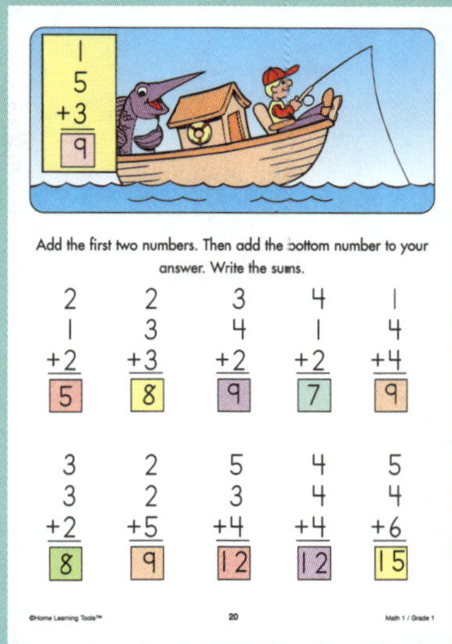
Add the first two numbers. Then add the bottom number to your answer. Write the sums.

2 1 +2 = 5	2 3 +3 = 8	3 4 +2 = 9	4 1 +2 = 7	1 4 +4 = 9
3 3 +2 = 8	2 2 +5 = 9	5 3 +4 = 12	4 4 +4 = 12	5 4 +6 = 15

page 21

Add or subtract to solve the problems. Write the answers.

23 −12 = 11	35 +42 = 77	61 +4 = 65	55 −22 = 33	45 −23 = 22
75 −34 = 41	32 +17 = 49	86 −25 = 61	44 −13 = 31	72 +27 = 99

page 22

Write the missing numbers.

5 +1 = 6	11 −4 = 7	7 +5 = 12	12 −6 = 6
5 +7 = 12	10 −6 = 4	8 +2 = 10	6 −4 = 2
4 5 +2 = 11	6 2 +5 = 13	2 5 +5 = 12	5 3 +7 = 15

page 23

Measure each thing. Circle the correct length on the ruler.

Circle the thing that is the longest.

page 24

The short hand tells the hour. Read each clock. Write the time.

3:00 5:00 8:00

10:00 1:00 6:00

page 25

The long hand tells the minutes. Read each clock. Write the time.

3:30 12:15 2:30

7:15 10:30 4:45

page 26

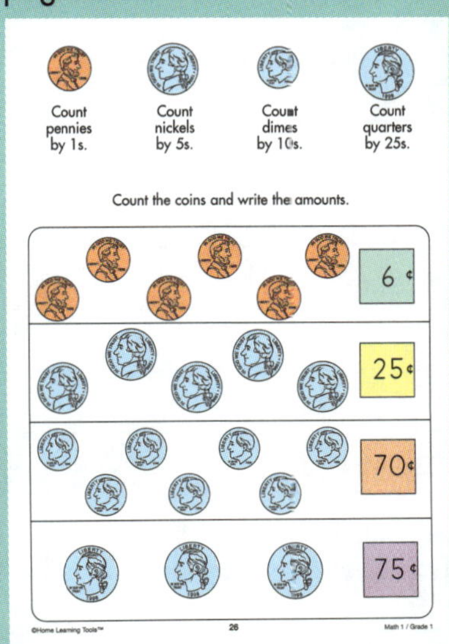
Count pennies by 1s. Count nickels by 5s. Count dimes by 10s. Count quarters by 25s.

Count the coins and write the amounts.

6¢

25¢

70¢

75¢

page 27

Circle the coins you need to buy each thing.

Answer Page

page 28

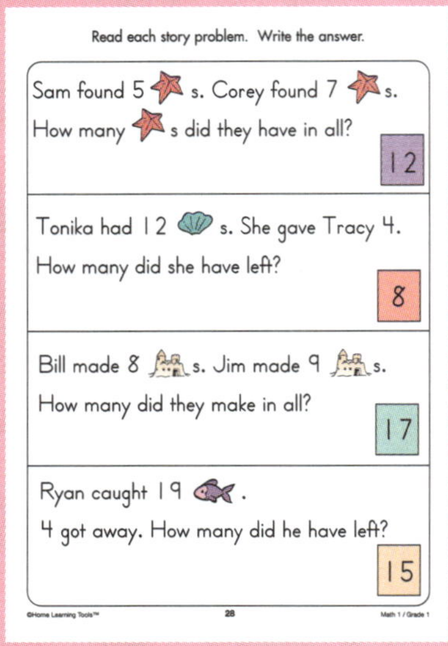
Read each story problem. Write the answer.

Sam found 5 s. Corey found 7 s.
How many s did they have in all? 12

Tonika had 12 s. She gave Tracy 4.
How many did she have left? 8

Bill made 8 s. Jim made 9 s.
How many did they make in all? 17

Ryan caught 19 .
4 got away. How many did he have left? 15

©Home Learning Tools™ 28 Math 1 / Grade 1

Your Letter Counts!

Home Learning Tools™ is proud to bring the highest level of quality education into your home. Please tell us about **YOURSELF** and **YOUR IDEAS** for our educational products!

You can write us at:

Home Learning Tools — P.O. Box 1823, Brentwood, TN 37024

Thank you and **CONGRATULATIONS** on your decision to learn the **Home Learning Tools** way.

I am most interested in teaching or learning about: ______________________________

____

I like **Home Learning Tools** because: ______________________________

____

I would like **Home Learning Tools** more if: ______________________________

____

Name ______________________________

Address ______________________________

Phone ______________________________ Age of Student ______________

10435MTH1 ______________________________ signature ______________ date